Burps

by Grace Hansen

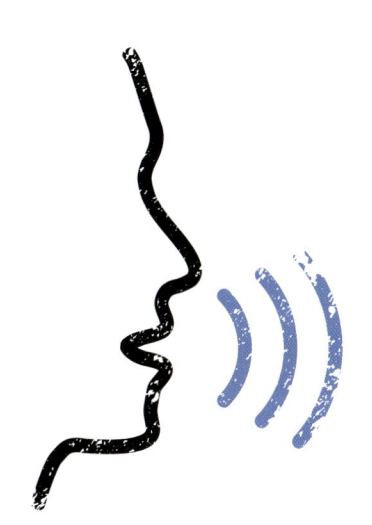

BEGINNING SCIENCE:
GROSS BODY FUNTIONS

Abdo Kids

Abdo Kids Jumbo is an Imprint of Abdo Kids
abdobooks.com

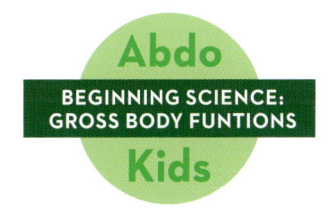

abdobooks.com

Published by Abdo Kids, a division of ABDO, P.O. Box 398166, Minneapolis, Minnesota 55439. Copyright © 2021 by Abdo Consulting Group, Inc. International copyrights reserved in all countries. No part of this book may be reproduced in any form without written permission from the publisher. Abdo Kids Jumbo™ is a trademark and logo of Abdo Kids.

Printed in the United States of America, North Mankato, Minnesota.

052020

092020

THIS BOOK CONTAINS RECYCLED MATERIALS

Photo Credits: iStock, Shutterstock

Production Contributors: Teddy Borth, Jennie Forsberg, Grace Hansen
Design Contributors: Dorothy Toth, Pakou Moua

Library of Congress Control Number: 2019956469
Publisher's Cataloging-in-Publication Data

Names: Hansen, Grace, author.

Title: Burps / by Grace Hansen

Description: Minneapolis, Minnesota : Abdo Kids, 2021 | Series: Beginning science: gross body functions | Includes online resources and index.

Identifiers: ISBN 9781098202361 (lib. bdg.) | ISBN 9781644943830 (pbk.) | ISBN 9781098203344 (ebook) | ISBN 9781098203832 (Read-to-Me ebook)

Subjects: LCSH: Human body--Juvenile literature. | Belching--Juvenile literature. | Digestion--Juvenile literature. | Excretion--Juvenile literature. | Hygiene--Juvenile literature.

Classification: DDC 612--dc23

Table of Contents

Loud & Stinky 4

Too Much Gas 10

What's That Smell? 16

Let's Review! 22

Glossary 23

Index 24

Abdo Kids Code 24

Loud & Stinky

Burps can sound and smell pretty gross. But they are doing an important job!

5

When we eat and drink, we swallow food and water. But we also swallow air.

7

Air contains gases like nitrogen and oxygen.

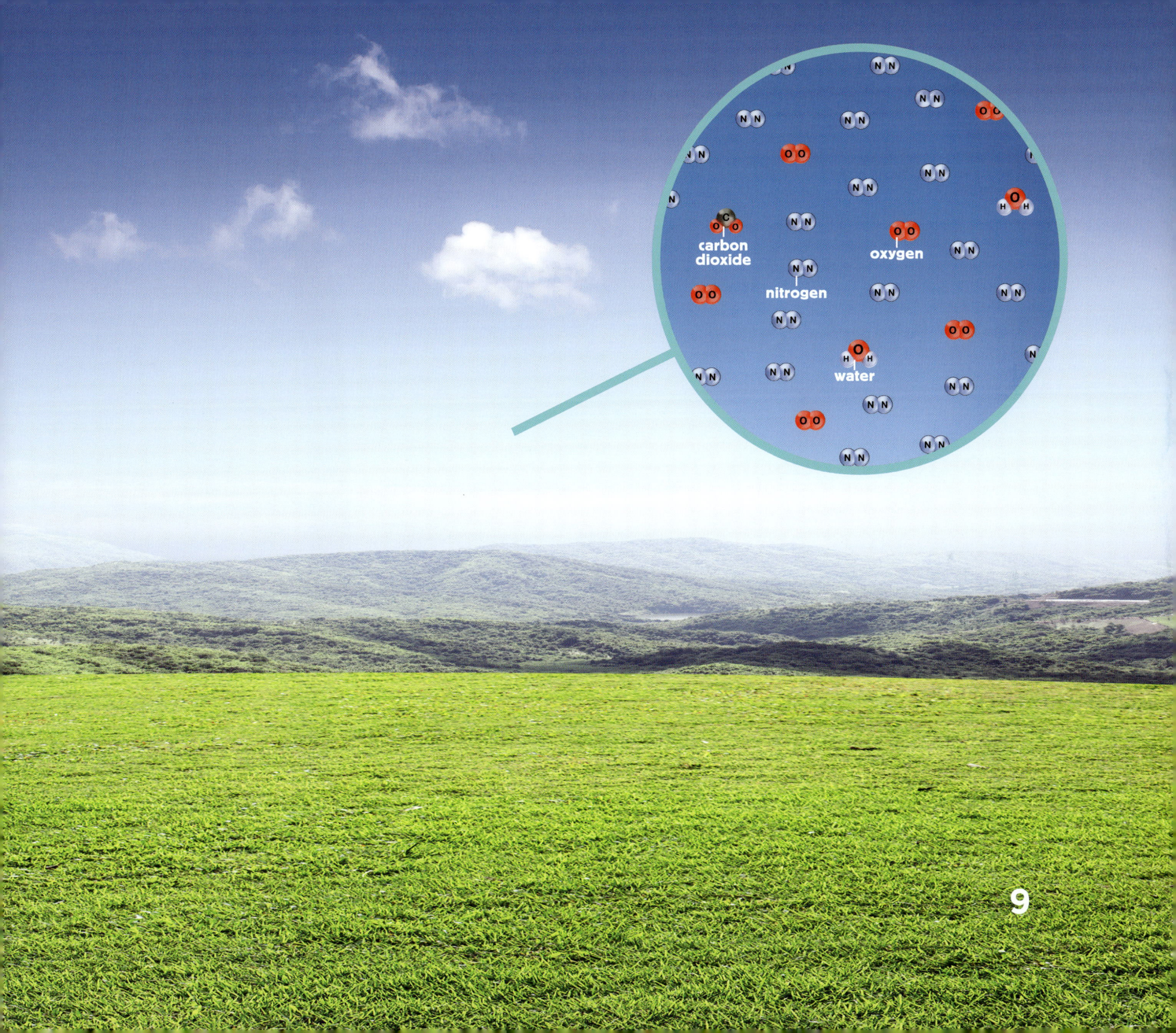

carbon dioxide

oxygen

nitrogen

water

Too Much Gas

Too much gas in the **esophagus** and stomach can be uncomfortable. Burping releases this **excess** gas.

esophagus

stomach

Fizzy drinks, like soda pop, can also cause us to burp. Soda pop is carbonated. It contains carbon dioxide.

13

Carbon dioxide is a gas. This gas is what gives a fizzy drink its bubbles!

15

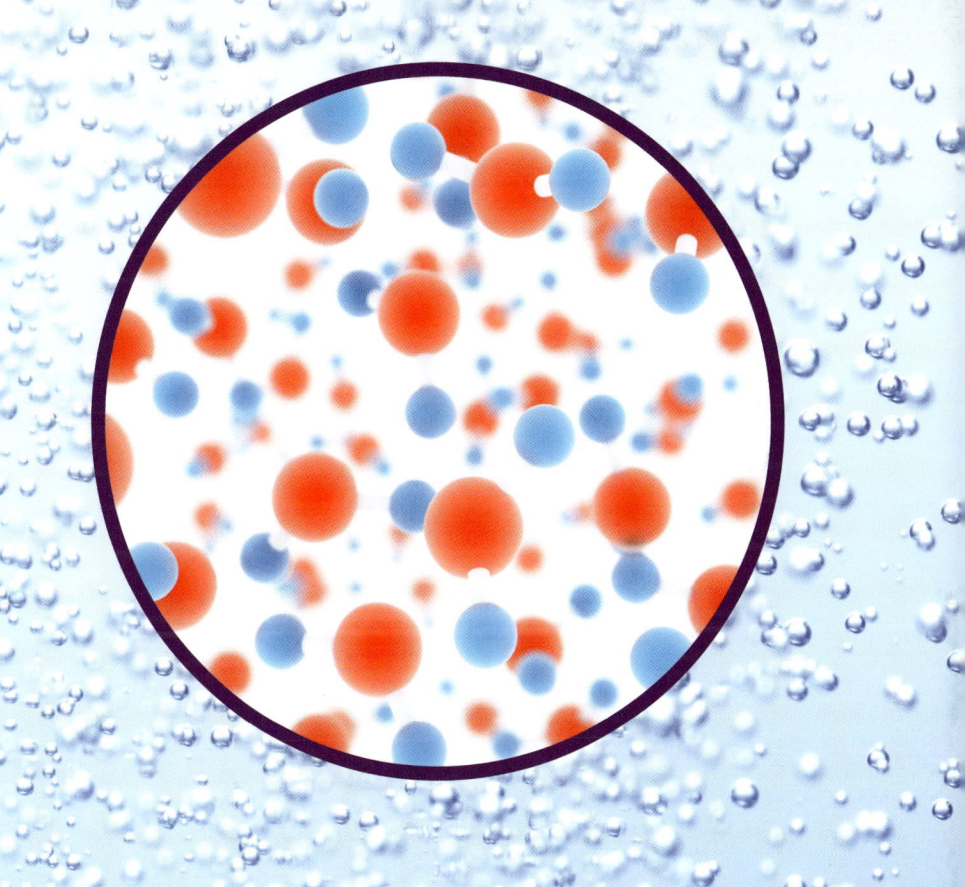

What's That Smell?

Sometimes burps smell really gross! This bad smell can be caused by certain things we eat and drink.

17

Bacteria in our **digestive tract** help break down food. Hydrogen sulfide can be made in the process. This gas smells like rotten eggs.

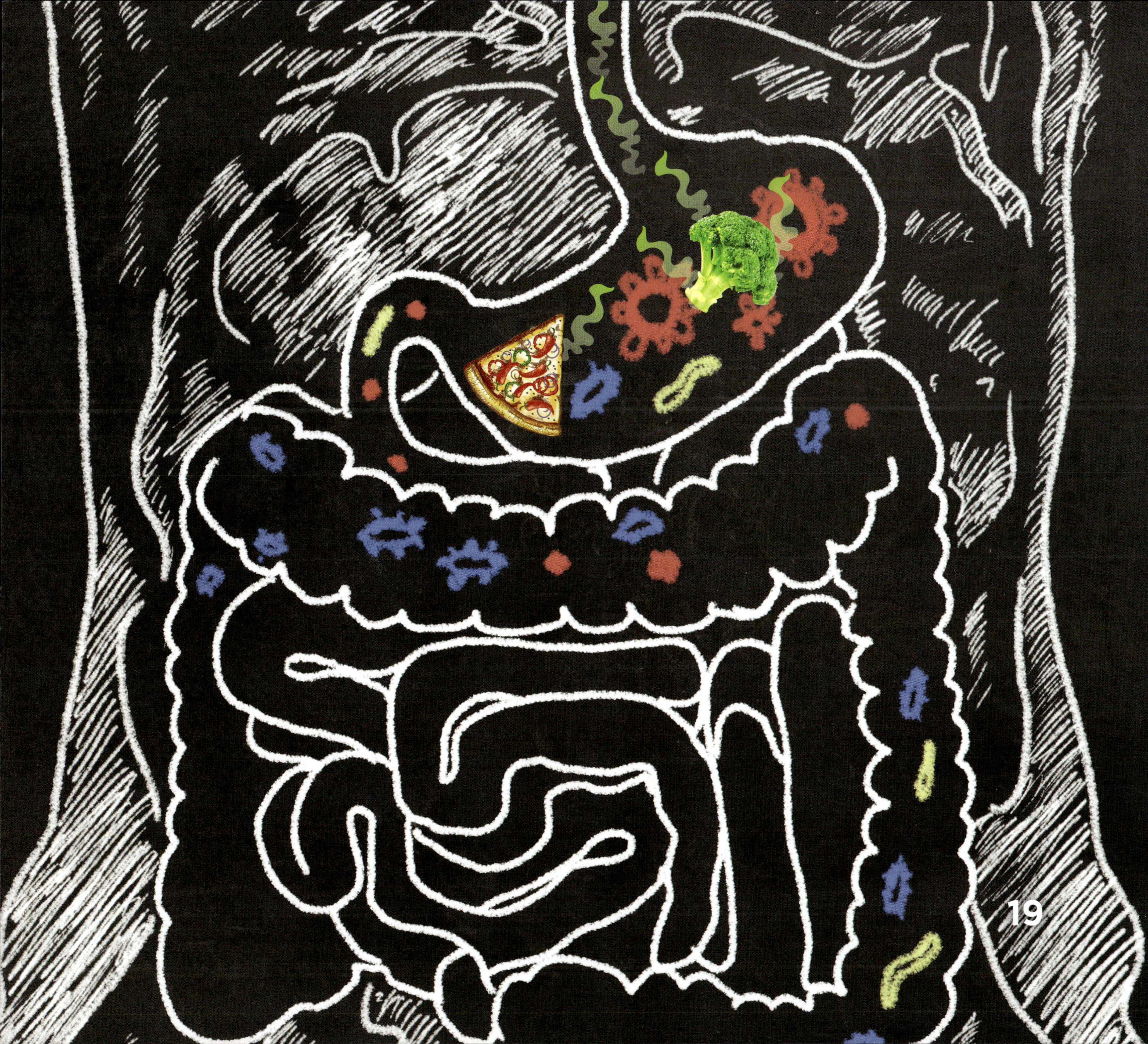

Everyone burps. But burping at certain times can be rude. Sometimes we cannot control when we burp. If you have to burp, it is always polite to say, "Excuse me!"

21

Let's Review!

- Burping is our body's way of releasing **excess** gas.

- Excess gas in the body can happen a few ways. Sometimes we swallow air when we eat or drink. Other times gas is created when **bacteria** break down food in our **digestive tract**.

- Carbonated drinks, like pop and fizzy water, can also create excess gas.

- Smelly burps are caused by the types of foods and liquids we eat and drink. Foods like cauliflower, broccoli, and dairy products can cause bad-smelling gas.

Glossary

bacteria – inside the human digestive tract, microscopic organisms that help in the digestion of food.

digestive tract – the parts of the body that work together to break down food so that it can be used by the body as energy. The human digestive tract, or system, includes the mouth, esophagus, stomach, and intestines.

esophagus – a tube that moves food and drink from the mouth to the stomach.

excess – an amount that is more than what is needed.

Index

air 6, 8

bacteria 18

carbonation 12

drinking 6, 16

eating 6, 16

esophagus 10

food 6, 16

gas 8, 10, 12, 14, 18

politeness 20

stomach 10

water 6

Abdo Kids ONLINE
FREE! ONLINE MULTIMEDIA RESOURCES

Visit **abdokids.com** to access crafts, games, videos, and more!

Use Abdo Kids code **BBK2361** or scan this QR code!